Bibliographic information published by the German National Library:

The German National Library lists this publication in the National Bibliography;
detailed bibliographic data are available on the Internet at http://dnb.dnb.de .

Imprint:

Copyright © 2017 GRIN Verlag, Open Publishing GmbH
Print and binding: Books on Demand GmbH, Norderstedt Germany
ISBN: 9783668460423

This book at GRIN:

http://www.grin.com/en/e-book/367553/yersinia-pestis-a-brief-overview-on-its-history-and-biology

Alexandria Villa

Yersinia Pestis. A Brief Overview on its History and Biology

GRIN Publishing

GRIN - Your knowledge has value

Since its foundation in 1998, GRIN has specialized in publishing academic texts by students, college teachers and other academics as e-book and printed book. The website www.grin.com is an ideal platform for presenting term papers, final papers, scientific essays, dissertations and specialist books.

Visit us on the internet:

http://www.grin.com/

http://www.facebook.com/grincom

http://www.twitter.com/grin_com

Alexandria Villa

Writing 122

21 March 2017

Yersinia pestis

Throughout history, mankind and diseases have coexisted, sometimes maintaining a delicate balance, and other times waging full out war upon each other. Some diseases only cause mild discomfort, like the many strains of the virus that causes the common cold, to the bacteria that lives inside of us that can potentially make us deathly ill, such as Escherichia coli, better known as E. coli.

Surviving and thriving with diseases become an integral part of life and living, even though this has also caused many history-changing moments as man has had to learn that, no matter how strong humanity thinks it is, there are things out there that he must bow to. One such disease was one that came in three different surges, decimating the world as it was then, shaping cultures and folklore, as well as beginning the move toward trying to identify what these invisible soldiers are and how to protect from all the malevolent 'germs'. This disease is formally known as *Yersinia pestis*, but most know it by its more common name, the Bubonic Plague, the Black Death, or the Black Plague.

One of the worst diseases to hit humanity was not something that comes from water, or even the air. This microorganism came directly from infected rats, gaining infectivity as it moved from rat to flea, from flea to human (Waterman). *Y. pestis* is unique for its origins, as it is one of the only major bacterium that does not depend on human beings as hosts for any gains, and, when 'processed' in just the right way, can become the initial ground for one of the most sinister bacterium to ever be introduced to mammals (Tucker).

Y. pestis, when looking at the organism through a microscope, is a rod-shaped bacterium. One of the most important aspects of the bacterium is that it has been documented to grow the best around 82 degrees Fahrenheit, but can survive and thrive in temperatures up to 104 degrees Fahrenheit (Perry and Fetherston). This, of course, means that this bacterium will find the human bodies temperature to be an optimal breeding ground. This would also mean that an infected person would have to have an uncomfortably high fever for it to begin killing of any bacterium that is within the body, and even then, the 104-degree temperature is only on the uncomfortable edge for the microorganism itself.

Infected rat blood, when ingested by fleas, is processed as any other blood meal, working through the digestive system inside of the abdomen of the flea. However, as the infection takes hold, the bacterium slowly grows into a large mass inside of the parasite. This blockage is made entirely of the bacterium. This mass begins to wreak havoc on the flea's ability to consume blood, resulting in the flea's inability to gain nutrients from its blood meals. This is an interesting attribute to *Y. pestis*, as the infection causes the flea to slowly starve to death. The flea begins to feed more often in a feeble attempt to survive, subsequently passing the disease on to those it feeds on at an even higher rate (Hinnebusch).

The transmission of the disease is also very interesting, as while it is found in various species of rats, it only begins to spread outside of the rat colonies once the fleas begin to move on to various other hosts. Rats have been found to work best as both vectors of the disease, as well as being the easiest to study, as they are easily containable in specific environments, and fleas are possibly the number one creature to feed upon the rat subjects. Studies have shown that there are over 1,500 species of fleas that can be possible vectors of the disease, but two have been instrumental to research for their ease in studies and the fact they are some of the most commonly found (Perry and Fetherston). These two are *Oropsylla montana*, a flea that primarily effects Ground Squirrels, and *Xenopsylla cheopis,* that feeds off the Oriental Rat. *X. cheopis* is the species of flea that is commonly associated with *Y. pestis* (Hinnebusch).

The differences between these two different flea species are what make them so susceptible for being the optimal carriers of the disease. Studies show that *O. montana,* once it succumbs to infection, is remarkably more infective during the first few days after contact with the bacterium, but it fades rather rapidly until the parasite either clears the bacterium or succumbs to it. *X. cheopis,* on the other hand, holds a pattern that is the exact opposite. As the infection rages within the parasite, the infectivity increases.

Studies have also shown that *O. montana* has an enlarged esophagus in comparison to *X. cheopis*, possibly allowing more of the bacterium to pass from the infected flea into the host. Infected fleas of both species were documents as having esophagus enlargement in both blocked and unblocked fleas (Hinnebusch).

Fleas play a major role in the spreading of the bacterium *Yersinia pestis*, transmitting it from rats and other rodents and vermin to other members of the mammal kingdom. Since fleas are hard to get rid of and are extremely mobile, it does not take long for them to pass on their diseases to human beings. The Bubonic Plague is one such disease that wreaked havoc on man for many centuries, reoccurring in the same areas multiple times over a substantial amount of time. *Y. pestis* is possibly one of the most ruthless bacterial infections, killing almost everything and everyone in a small amount of time, including the fleas that carry it from host to host.

One such disease came in three different surges, decimating the foundations of the known world as it was, shaping cultures and folklore, and kick-starting the move for humanity to begin the long trek to discover the nature of disease and the malevolent organisms that caused them. This disease is formally known as *Yersinia pestis,* but most of the world knows it as The Black Plague, something that many people do not realize still effects humans to this day.

Yersinia pestis has been around for a very long time. Many believe that the disease had been lurking in Asia for centuries before the major outbreaks occurred. It is not a completely proven fact, but most historians believe that the Black Plague first came to Europe in the years sometime before 700 AD. It is widely believed that up to 40% of Constantinople's population was decimated in this time, but records are scarce (Ray). Due to the lack of found records, it is impossible to say how many lost their lives, or the precise details behind the causes of the outbreak. What brought it, and how quickly it filtered through the population remains a mystery.

Much more information is known about the second major Bubonic Plague outbreak. Trade routes were opening up and becoming popular during this time of human history, allowing goods to be transported from one continent to another. However, there was more than just shipments of goods moving from place to place with the ships. It is thought that a fleet of twelve Genoese ships brought the disease from Asia to Sicilian port of Messina, because the ships of that time were known to be infested with vermin, making it completely likely that, while the ships were at a different port, infected rats were able to make their way onboard. After they make their way onto these floating asylums, the rats slowly spread the disease amongst themselves before the disease jumped from rats and infected fleas, all the way to humans. Reports claim that when the ships finally docked at Messina, most of the sailors on board were found dead. The few remaining survivors were riddled with sores and large, black boils that oozed pus and blood, which are known now as the infamous signs of *Yersinia pestis* infection (History.com).

Of course, it was these overt symptoms that gave the disease its spine-chilling and simplistic name. The Sicilians attempted to send the ships, and the strange and foreign disease they carried, back into the harbor, but it was already far too late. The infected rats and fleas had already infiltrated the port. These infected creatures had already begun to spread their horrible disease onward. Over the next five years, the Bubonic Plague would proceed to kill over 20 million people in Europe, almost one third of the continent's population (History.com).

The third outbreak was potentially even more harsh than the second one, but this particular epidemic was far less famous. This third outbreak was in China in 1855. While this epidemic only started out killing tens of thousands, the world was far more connected than it had ever been before in the history of humanity. The Bubonic Plague is thought to have originated in China's Yunnan Province, and by the late 19th century had spread to Hong Kong and Guangzhou. From there, it seems as though India was the next to experience the horrendous effects of the disease. From there, it spread to the rest of the world (Sanburn).

The World Health Organization shows that 12 million people succumbed to the Black Plague during this epidemic. Luckily, though, at the time of this pandemic, doctors and scientists were capable of finding out various bits of information to work on identifying and treating such a horrible disease (Sanburn). This outbreak did not end until the early 19th century. Since the 1855 outbreak was seemingly so sporadic, there were relatively large periods of time for scientists to isolate and study the bacteria that caused the outbreaks.

In 1892, a man named Alexandre Yersin entered the picture. He had already done extensive research on microbiology, helping to discover and isolate the toxin secreted by diphtheria bacterium. Yersin was sent to Hong Kong in 1894, where he and a colleague

independently discovered the bacterial agent behind the Bubonic Plague. The bacterium was

originally named *Pasteurella pestis*, after famed scientist Louis Pasteur, but later the name was

changed to commemorate the discoverer himself (Encyclopædia Britannica).

This identification and discovery was a major achievement, even on a global scale.

Knowing the nature of the cause of one of the worst diseases in human history was an

extraordinary component of learning how to combat the illness and lessen the impact it had on

society. However, this disease is still around, taking hold of people every so often and showing

the world that, despite our best efforts, some diseases may be contained, but will never fully be

eradicated.

Yersinia pestis is one of the most violent diseases known to man. This disease is caused

by a bacterial infection that manifests itself after an incubation period of three to seven days.

After that, the symptoms all depend on the way the victim contracted the disease.

One way that the disease is contracted is through the respiratory system. When the

disease is transmitted this way, it is called Pneumonic Plague. This way usually occurs when a

victim breathes in the bacteria, whether it be from another infected victim, like mucus released

from a cough or sneeze, or from inhaling the bacteria through dealing with infected animal feces

(Admin).

Either way, it leads to the bacteria reproducing within the lungs. Pneumonic Plague

begins with the same symptoms as normal Pneumonia, with fevers, harsh, rattling coughs, fluid

build-up in the lungs, and so forth. However, after three to five days, the victim begins to cough

up bloody mucus, and eventually coughs up lung tissue, blood, and bacteria. This usually results

in death around day five, if left untreated in the very early phases. This is the most infective form of the plague, and usually the most infective when it comes to human-human transmission (MedicineNet Staff).

The second form of the Bubonic Plague is called Septicemic Plague. As the names have a tendency to imply, different forms mimic the beginning phases of other diseases. If victims have an open wound, whether it be a small cut, large gash, or a scratched-open flea bite, they will be more likely to fall into this category type of plague. The victim contracts the disease through open wounds, where it spreads through the body just like septicemia. Some of the symptoms include fever, chills, vomiting, extreme weakness, and potentially developing large bruises on the skin, revealing leaking capillaries (Admin). This form is only slightly less contagious than the other forms. As the bacteria take hold inside of the victim's body, it also pervades the respiratory system, allowing it to spread like the Pneumonic Plague form.

Interestingly enough, however, victims of Septicemic Plague do not develop the tell-tale sign of buboes. Buboes are the large balls that develop on the armpits, groin, and neck of those who have contracted certain forms of the Bubonic Plague. This form of the plague usually kills untreated hosts within a week of contracting the bacterial infection (Center for Disease Control).

The third form of this plague is full-fledged Bubonic Plague. This form manifests itself as large, painful swelling buboes on the site where the lymph nodes are. The bacteria enter the body through an infected bite, and basically congregate in the nearest lymph node.

As the victim's immune system attempts to fight off the infection, the bacteria begin to spread quickly. This causes all other lymph nodes to eventually become swollen. As it rages on, the buboes grow and darken, causing the appearance that gave the disease its names: Black Plague and Black Death. This is the form that most people think of when they hear 'Bubonic

Plague'. It could also potentially be the most painful way to pass, as the immune system is attacked directly and acts as the main force to spread it through the body (Admin). Your own body becomes your worst enemy.

Because of the nature of the symptoms, many doctors have been accidentally guilty of misdiagnosing the disease. This causes many cases to go improperly treated, as well as untreated completely in some cases of Pneumonic plague (CDC). This is what makes education so important. For this disease, the problem is not so much a case of the cause being little known or understood, but more a case of misguided comfortableness because many people out there think that this disease is something of the past, and not relevant in the modern day.

Doctors seem to have a tendency of not looking into their patient's symptoms deeply enough, at least not until it is too late. *Y. pestis* has an average incubation period of three to seven days, and when the first symptoms show up, they can look like a variety of different conditions with different causes. The main issue with this is, if the disease is left improperly treated, it will spread, and potentially harm or kill others as well.

The first step for treatment is proper diagnosis, just like any other disease. A medical professional will review the victim's obvious symptoms, looking for things like sudden onset fever, chills, and headaches. They will also look for animal or insect bites, as it is through these avenues that infection is more likely to occur. If there is suspicion of infection after this, then samples of blood, sputum or mucus, and lymph node aspirate are taken and sent to a lab. At this point, an antibiotic could be given as presumptive treatment (CDC).

Once the lab gets the samples, a series of preliminary tests are conducted. These can be done and analyzed within two hours. If there is potential cause, however, the labs will conduct more comprehensive tests, with the entire process taking up to two days. By then, in most cases an antibiotic will be introduced (CDC).

Once the case is proven to be caused by the Bubonic Plague, the patient is then hospitalized and sent into isolation to prevent spreading the infection. From there, a series of antibiotics are given to the patient. Common antibiotics used for this purpose and are known to be remarkably effective are Gentamycin and Streptomycin, with Tetracyclines and chloramphenicol added to the list in the early 2000's. In many cases, anyone that the victim had come into contact with will be traced, found, and evaluated in order to keep the impact at an absolute minimum (Schoenstadt).

Overall, this disease is not untreatable like it was in the Middle ages. There have been so many advancements in this particular field, and all of that knowledge has come a long way to helping humanity as a whole. All it takes is the understanding of the general public. If everyone, whether it be the lonely hermit who watches the weather from the hillside or the angry lady on the bus who is always telling people what exactly what she thinks, could recognize the potential risks then there would not be the threat of preventable outbreaks. In the modern day, the Bubonic Plague only has a 50% mortality rate if caught early enough. If caught too late, that mortality rate rises to 90%, all paired with the possibility of uncontrollable spreading (Schoenstadt). In this case, like so many others, education and prevention are the best ways to combat illness.

Works Cited

Admin, – By. "Symptoms and Treatment." *Symptoms and Treatment RSS*. Symptoms Treatment

Organization, 18 Jan. 2012. Web. 01 Mar. 2017.

This is an in-depth article on the various symptoms of the Black Plague as well as the

diagnosis and treatment of the disease.

CDC, "Frequently Asked Questions: Plague." Centers for Disease Control and Prevention.

Centers for Disease Control and Prevention, 14 Sept. 2015. Web. 13 Feb. 2017.

This article was compiled by the CDC. It gives generalized and sometimes detailed

information about the nature of the Bubonic, or Black, Plague. It was designed to keep

people informed in a concise manner, with an emphasis on Frequently Asked Questions.

Center for Disease Control and Prevention. "Symptoms." *Centers for Disease Control and*

Prevention. Centers for Disease Control and Prevention, 14 Sept. 2015. Web. 01 Mar.

2017.

CDC quick fact guide to the attributes of the different forms of the Black Plague.

Durham, David P. and Elizabeth A. Casman. "Threshold Conditions for the Persistence of Plague

Transmission in Urban Rats." *Risk Analysis: An International Journal*, vol. 29, no. 12,

Dec. 2009, pp. 1655-1663. EBSCO*host*, doi:10.1111/j.1539-6924.2009.01309.

This report goes into extreme detail in an experiment involving rats and the transmission

of *Y. pestis* if introduced to an urban population.

The Editors of Encyclopædia Britannica. "Alexandre Yersin." *Encyclopædia Britannica.*

Encyclopædia Britannica, Inc., 16 Dec. 2009. Web. 01 Mar. 2017.

This is a comprehensive entry on the nature of Alexandre Yersin, the founder of *Y. pestis.* It also details his life, education, family, history, and accomplishments.

Hinnebusch, B. Joseph, et al. "Comparative Ability of Oropsylla Montana And Xenopsylla

Cheopis Fleas To Transmit Yersinia Pestis By Two Different Mechanisms. *Plos*

Neglected Tropical Diseases 11.1 (2017): 1-15, *Academic Search Complete.* Web. 23

Jan. 20017.

This in-depth article tells about the nature of two species of fleas, and gives a

comprehensive analysis of how *Yersinia pestis* effects the internal structures of fleas as

well as how well each species of flea transmits the bacteria to another host.

History.com Staff. "Black Death." *History.com.* A&E Television Networks, 20 June 2010. Web.

01 Mar. 2017.

This article from History.com gives comprehensive information on the three major

epidemics that plagued humanity throughout time.

MedicineNet Staff. "Plague Symptoms, Treatment, Causes - Septicemic Plague." *MedicineNet.*

MedicineNet.com, 03 Sept. 2009. Web. 01 Mar. 2017.

This article covers most of the information about the Bubonic Plague, including

information on the bacterial aspect of it, major symptoms, as well as information

reguarding diagnosis.

Perry, R.D., and Jaqueline D. Fetherstone. "R D Perry." *Yersinia pestis – etiologic Agent of*

Plague. American Society of Microbiology, 01 Jan. 1997. Web. 25 Jan. 2017.

This article delves into the history of the Bubonic Plague, detailing the nature of each of

the three major outbreaks and listing detailed information about the nature of the plague.

Ray, Michelle. "The Black Plague." *Black Death.* Temple University, 13 Apr. 2011. Web. 01

Mar. 2017. This website gives various bits of information regarding the plague. It

includes information on its history, the nature of the disease, and other facts relevant to

the subject.

Sanburn, Josh. "Top 10 Terrible Epidemics." *Time.* Time Inc., 26 Oct. 2010. Web. 01 Mar. 2017.

This article focuses on ten of the worst epidemics that has ever hit humanity. One of the

entries is on the Black Plague and gives facts and information on the subject.

Schoenstadt, MD Arthur. "Bubonic Plague Treatment." *EMedTV: Health Information Brought*

To Life. EMedTV, 26 Jan. 2017. Web. 21 Mar. 2017.

This article gives detailed information on the evaluation and treatment of the Bubonic

Plague in an easily accessible and ascertainable online format.

Tucker, Brandon. "Yersinia Pestis." *Yersinia pestis.* Missouri University of Science and

Technology, 16 Feb. 2010. Web. 28 Jan. 2017.

This site gives an overview on the structures and symptoms of *Y. pestis.* It includes

diagrams and brief information on transmission and structures.

Waterman, Hillary. "A Profile of Plague." *Discover* 37.8 (2016): 78. *MasterFILE Premier.* Web.

28 Jan. 2017.

This magazine article gave information on various aspects of the Black Plague, like how

the bacteria is transmitted as well as listing facts about the prevalence of the disease

today.

YOUR KNOWLEDGE HAS VALUE

- We will publish your bachelor's and master's thesis, essays and papers

- Your own eBook and book - sold worldwide in all relevant shops

- Earn money with each sale

Upload your text at www.GRIN.com and publish for free